SWORDTAILS FISH PET FOR NEWBIES

The Indispensable Handbook For Novices

MANUEL S. ANDERSON

Copyright © 2024 MANUEL S. ANDERSON

All rights reserved. No part of this publication may be reproduced, distributed, or transmitted in any form or by any means, including photocopying, recording, or other electronic or mechanical methods, without the prior written permission of the publisher, except in the case of brief quotations embodied in critical reviews and certain other noncommercial uses permitted by copyright law.

TABLE OF CONTENTS

TABLE OF CONTENTS……………...…2-3

INTRODUCTION TO SWORDTAIL FISH.4-7

CHAPTER 1: Overview of Swordtails……8-18

History And Origins of Swordtail Fish…..12-15

Why Choose Swordtails as Pets……..…...15-18

CHAPTER 2: Swordtail Varieties………..19-34

Different Types and Colors of Swordtail Fish……………………………………….22-26

Characteristics and Features of Swordtail Fish……………………………………..26-30

Selecting the Right Swordtail for You…...30-34

CHAPTER 3: Setting Up the Ideal Swordtail Aquarium………………………..…..35-51

Tank Size and Setup for Swordtail Fish….39-43

Water Parameters and Filtration for Swordtail Fish…………………………………….44-47

Substrate, Decorations, and Plants for Swordtail Fish Aquarium…………………………48-51

CHAPTER 4: Swordtail Fish Care Guide..52-69

Feeding Swordtails............................56-60

Water Maintenance and Testing for Swordtail Fish...60-64

Handling Health Issues.......................64-69

CHAPTER 5: Swordtail Breeding..........70-87

Breeding Basics................................74-78

Breeding Tank Setup..........................78-82

Caring for Swordtail Fry.....................83-87

CHAPTER 6: Compatibility and Tankmates.....................................88-102

Suitable Tankmates for Swordtails.........91-94

Potential Compatibility Issues..............95-98

Creating a Harmonious Community Tank...98-102

CHAPTER 7: Tips for Success with Swordtails..................................107-110

Common Mistakes to Avoid..............110-114

CONCLUSION.............................115-116

INTRODUCTION TO SWORDTAIL FISH

Origins and Natural Habitat:

Swordtail fish, scientifically known as Xiphophorus hellerii, originate from the freshwater streams and rivers of Central America, particularly Mexico and Honduras. In their natural habitat, Swordtails thrive in warm waters with moderate to high levels of vegetation and ample hiding spots among rocks and aquatic plants.

Physical Characteristics:

One of the most distinguishing features of Swordtail fish is the elongated, sword-like extension on the lower part of their tails, which is more prominent in males. These fish exhibit a variety of colors and patterns, ranging from solid hues of red, orange, yellow, and green to intricate combinations of spots, stripes, and iridescence.

Behavior and Social Structure:

Swordtails are known for their active and social nature, often seen swimming around the middle and upper levels of the aquarium. They are peaceful community fish that generally get along well with other non-aggressive species. However, males can sometimes display territorial behavior, especially towards other males or during mating.

Diet and Feeding Habits:

In the wild, Swordtail fish are omnivores, feeding on a varied diet of small invertebrates, algae, and plant matter. In captivity, they readily accept a range of commercial flake, pellet, and freeze-dried foods formulated for tropical fish. Supplementing their diet with fresh vegetables like blanched spinach or zucchini can help mimic their natural feeding habits and promote optimal health.

Reproduction and Breeding:

Swordtails are prolific breeders, with females capable of producing multiple batches of fry throughout their lifespan. Breeding Swordtails in the aquarium can be a rewarding experience, but it's essential to provide suitable breeding conditions, such as ample hiding places for fry and a separate breeding tank to protect them from predators.

Care and Maintenance:

Maintaining optimal water quality is crucial for the health and well-being of Swordtail fish. They prefer slightly alkaline water with a pH range between 7.0 to 8.0 and temperatures ranging from 72°F to 82°F (22°C to 28°C). Regular water changes, filtration, and monitoring of ammonia, nitrite, and nitrate levels are essential for keeping Swordtails thriving in the aquarium.

Therefore, Swordtail fish are a fantastic addition to any freshwater aquarium, offering beauty, personality, and ease of care for both beginner and experienced hobbyists alike. By understanding their natural habitat, behavior, dietary needs, and breeding tendencies, aquarists can create a thriving environment that showcases the beauty and vibrancy of these captivating fish. Whether kept in a community tank or as part of a dedicated breeding project, Swordtails never fail to impress with their striking colors and dynamic personalities.

CHAPTER 1: Overview of Swordtails

Swordtail fish, scientifically known as Xiphophorus hellerii, are a popular freshwater species among aquarium enthusiasts worldwide. Here's an overview highlighting their key features:

Physical Appearance:

Swordtails are known for the elongated, sword-like extension on the lower part of their tails, particularly prominent in males.

They come in a wide range of colors, including red, orange, yellow, green, and various combinations of these hues.

Some swordtail varieties also exhibit unique patterns, such as spots, stripes, and iridescence, adding to their visual appeal.

Behavior and Temperament:

Swordtails are active and social fish, often seen swimming throughout the middle and upper levels of the aquarium.

They generally have peaceful temperaments and can coexist with a variety of non-aggressive tankmates.

Males may occasionally exhibit territorial behavior, especially during mating or when housed with other males.

Natural Habitat:

Swordtails originate from the freshwater streams and rivers of Central America, particularly Mexico and Honduras.

In their natural habitat, they prefer warm waters with moderate to high levels of vegetation and ample hiding spots among rocks and aquatic plants.

Dietary Needs:

Swordtails are omnivores, feeding on a varied diet of small invertebrates, algae, and plant matter in the wild.

In captivity, they readily accept commercial flake, pellet, and freeze-dried foods formulated for tropical fish.

Supplementing their diet with fresh vegetables like blanched spinach or zucchini can help mimic their natural feeding habits and promote optimal health.

Breeding Behavior:

Swordtails are prolific breeders, with females capable of producing multiple batches of fry throughout their lifespan.

Breeding swordtails in the aquarium can be a rewarding experience, but it's essential to provide suitable breeding conditions, such as ample hiding places for fry and a separate breeding tank to protect them from predators.

Care and Maintenance:

Maintaining optimal water quality is crucial for the health and well-being of swordtail fish.

They prefer slightly alkaline water with a pH range between 7.0 to 8.0 and temperatures ranging from 72°F to 82°F (22°C to 28°C).

Regular water changes, filtration, and monitoring of ammonia, nitrite, and nitrate levels are essential for keeping swordtails thriving in the aquarium.

Therefore, Swordtail fish are prized for their striking appearance, active behavior, and ease of care, making them a popular choice for beginner and experienced aquarists alike. By understanding their natural habitat, dietary needs, breeding tendencies, and care requirements, fishkeepers can create a thriving environment that showcases the beauty and vibrancy of these captivating freshwater fish.

History And Origins of Swordtail Fish

Swordtail fish (Xiphophorus hellerii) have a fascinating history and origin story that dates back to their native habitats in Central America. Here's a brief overview:

Native Habitat:
Swordtails are indigenous to the freshwater streams and rivers of Central America, primarily found in regions of Mexico and Honduras. In these habitats, they inhabit warm, slow-moving waters with abundant vegetation, rocks, and submerged roots, providing ample hiding spots and natural foraging grounds.

Discovery and Naming:
The scientific classification of swordtail fish, Xiphophorus hellerii, was first documented by the German zoologist and physician Johann

Jakob Heckel in the mid-19th century. The genus name "Xiphophorus" is derived from Greek words meaning "sword bearer," referring to the distinctive sword-like extension on the lower part of their tails. The species epithet "hellerii" honors the German naturalist Karl Bartholomäus Heller, who collected specimens in Mexico.

Domestication and Aquarium Trade:
Swordtails gained popularity in the aquarium hobby during the early 20th century, primarily due to their vibrant colors, distinctive appearance, and ease of care. Breeders and hobbyists began selectively breeding swordtails to enhance their coloration and finnage, resulting in a wide array of unique varieties available in the aquarium trade today.

Role in Scientific Research:

Swordtail fish have also played a significant role in scientific research, particularly in the fields of genetics, evolution, and cancer biology. Their relatively short generation time, large brood sizes, and genetic variability make them ideal subjects for studying various biological processes and phenomena.

Conservation Status:

While swordtail fish are not currently listed as endangered, their natural habitats face threats from habitat destruction, pollution, and invasive species. Conservation efforts are underway to preserve the biodiversity of Central American freshwater ecosystems and protect native fish species like the swordtail.

Therefore, the history and origins of swordtail fish showcase their journey from their native habitats in Central America to becoming popular and beloved aquarium inhabitants worldwide. Through centuries of exploration, scientific study, and selective breeding, swordtails have evolved into the vibrant and diverse species cherished by aquarists today. Understanding their history helps us appreciate the cultural and ecological significance of these fascinating freshwater fish.

Why Choose Swordtails as Pets

Swordtail fish (Xiphophorus hellerii) are an excellent choice for aquarium enthusiasts for several compelling reasons:

Vibrant Colors and Unique Appearance: Swordtails come in a wide array of colors, ranging from bold reds and oranges to subtle greens and blues.

Their distinctive sword-like extension on the lower part of their tails adds visual interest and makes them stand out in any aquarium.

Ease of Care:

Swordtails are relatively hardy and adaptable fish, making them suitable for beginner aquarists.

They can tolerate a range of water conditions and do well in community tanks with compatible tank mates.

Active and Social Behavior:

Swordtails are active swimmers that add energy and movement to any aquarium.

They exhibit social behaviors and can coexist peacefully with a variety of other fish species, enhancing the dynamics of the tank.

Breeding Potential:

Swordtails are prolific breeders, making them an excellent choice for hobbyists interested in breeding and raising fish.

Breeding swordtails can be a rewarding experience, as it allows enthusiasts to observe the fascinating process of reproduction and raise fry into healthy adults.

Educational and Scientific Value:

Swordtail fish have been subjects of scientific research due to their genetic variability and relevance to studies in genetics, evolution, and cancer biology.

Keeping swordtails as pets provides an opportunity for hands-on learning and observation of natural behaviors in a controlled environment.

Contribution to Conservation:

By keeping swordtails in captivity, aquarists

contribute to the conservation of freshwater fish species by supporting responsible breeding practices and reducing demand for wild-caught specimens.

Participating in breeding programs and sharing knowledge within the hobbyist community helps promote awareness and appreciation for the conservation of aquatic ecosystems.

Therefore, Swordtail fish offer a combination of aesthetic appeal, ease of care, and educational value that makes them an attractive choice for aquarium enthusiasts of all levels. Whether you're drawn to their vibrant colors, intrigued by their breeding behaviors, or interested in their scientific significance, swordtails have something to offer for every fishkeeper. With proper care and attention, swordtails can thrive in the aquarium and bring joy to their keepers for years to come.

CHAPTER 2: Swordtail Varieties

Swordtail fish (Xiphophorus hellerii) come in a diverse range of varieties, each with its own unique colors, patterns, and finnage. Here are some popular swordtail varieties commonly found in the aquarium trade:

Red Swordtail: This variety features vibrant red coloration throughout its body, fins, and tail. The sword extension on males is often red, creating a striking contrast against the rest of the body.

Black Swordtail: Characterized by deep black coloration, black swordtails are sleek and elegant, with a contrasting sword extension that adds visual interest. Some may have hints of other colors, such as red or orange, on their fins.

Marigold Swordtail: Marigold swordtails showcase bright yellow or orange hues, reminiscent of the petals of a marigold flower. Their swords may vary in color, ranging from black to red or even yellow.

Pineapple Swordtail: Pineapple swordtails exhibit a mosaic pattern of yellow, orange, and black markings, resembling the skin of a pineapple. Their swords may be solid black or have patterns that mirror the body coloration.

Tuxedo Swordtail: Tuxedo swordtails feature a two-tone coloration, with a darker body and contrasting lighter fins. The sword extension on males is typically black, creating a "tuxedo" effect.

Neon Swordtail: Neon swordtails are known for their bright and fluorescent colors, including neon blue, green, and pink. Their swords may be transparent or match the body coloration.

Wagtail Swordtail: Wagtail swordtails have a distinctive upward curvature in their tail fins, resembling the tail of a wagging dog. This variety comes in various colors, including red, black, and combinations of both.

Lyretail Swordtail: Lyretail swordtails have elongated and flowing tail fins that resemble the shape of a lyre instrument. This variety adds elegance and grace to any aquarium with its unique finnage.

Half Black Swordtail: Half black swordtails have a predominantly black coloration on the posterior half of their bodies, gradually fading to lighter colors towards the front. The sword extension may also be black or have a contrasting color.

Albino Swordtail: Albino swordtails lack melanin pigment, resulting in a pale pink or white body with red eyes. Their fins and swords may exhibit translucent or pale coloration.

These are just a few examples of the diverse swordtail varieties available in the aquarium hobby. With selective breeding and hybridization, enthusiasts continue to create new and unique swordtail strains, adding to the beauty and fascination of this beloved freshwater fish species.

Different Types and Colors of Swordtail Fish

Swordtail fish (Xiphophorus hellerii) are known for their diverse range of types and colors, making them a popular choice among aquarium enthusiasts. Here are some of the different types and colors of swordtail fish commonly found in the aquarium hobby:

Solid Color Varieties:

Red Swordtail: Featuring vibrant red coloration throughout the body, fins, and tail.

Black Swordtail: Characterized by deep black coloration, sometimes with hints of other colors on the fins.

Yellow Swordtail: Displaying bright yellow coloration, often with variations in intensity and shading.

Patterned Varieties:

Pineapple Swordtail: Exhibiting a mosaic pattern of yellow, orange, and black markings, resembling the skin of a pineapple.

Tuxedo Swordtail: Having a two-tone coloration, with a darker body and contrasting lighter fins, resembling a tuxedo.

Wagtail Swordtail: Featuring an upward curvature in the tail fins, resembling the wagging tail of a dog, often with red or black coloration.

Bi-Color Varieties:

Half Black Swordtail: Having a predominantly black coloration on the posterior half of the body, gradually fading to lighter colors towards the front.

Marigold Swordtail: Showcasing bright yellow or orange hues, often with a contrasting sword extension.

Albino and Leucistic Varieties:

Albino Swordtail: Lacking melanin pigment, resulting in a pale pink or white body with red eyes, often with translucent fins.

Leucistic Swordtail: Similar to albino swordtails but may have subtle differences in coloration and eye pigmentation.

Fancy and Hybrid Varieties:

Lyretail Swordtail: Featuring elongated and flowing tail fins resembling the shape of a lyre instrument, available in various colors and patterns.

Neon Swordtail: Displaying bright and fluorescent colors, including neon blue, green, and pink, often with transparent or translucent fins.

Longfin Varieties:

Longfin Swordtail: Having elongated and flowing fins, adding elegance and grace to the fish, available in a variety of colors and patterns.

Each type and color variation of swordtail fish adds beauty and diversity to the aquarium hobby. With selective breeding and hybridization, enthusiasts continue to create new and unique swordtail strains, further enhancing the appeal of this beloved freshwater fish species.

Characteristics and Features of Swordtail Fish

Swordtail fish (Xiphophorus hellerii) possess a range of characteristics and features that make them fascinating and attractive additions to freshwater aquariums. Here are some key traits:

Sword-Like Extension:
One of the most distinctive features of swordtail fish is the elongated, sword-like extension on the lower part of their tails, particularly prominent in males.

This extension varies in length and shape, adding visual interest and enhancing the fish's overall appearance.

Vibrant Coloration:

Swordtails come in a wide array of vibrant colors, including red, orange, yellow, green, blue, and various combinations thereof.

Some swordtail varieties also exhibit unique patterns, such as spots, stripes, and iridescence, adding to their visual appeal.

Active Behavior:

Swordtails are known for their active and energetic swimming behavior, often exploring all levels of the aquarium.

They enjoy ample space to swim and appreciate environments with plenty of plants and hiding spots to mimic their natural habitat.

Peaceful Temperament:

Swordtail fish generally have peaceful temperaments, making them suitable for community aquariums with compatible tankmates.

While males may exhibit occasional territorial behavior, especially during mating, they are typically not aggressive towards other fish species.

Hardiness and Adaptability:

Swordtails are relatively hardy and adaptable fish, capable of tolerating a range of water conditions and environmental parameters.

They can thrive in a variety of setups, including freshwater community tanks and planted aquariums, provided that water quality is maintained.

Prolific Breeding:

Swordtails are prolific breeders, with females capable of producing multiple batches of fry throughout their lifespan.

Breeding swordtails can be a rewarding experience for hobbyists, but it's essential to provide suitable breeding conditions and manage fry populations.

Educational Value:

Keeping swordtail fish in the aquarium provides an opportunity for hands-on learning and observation of natural behaviors in a controlled environment.

They can serve as educational tools for teaching concepts related to genetics, evolution, and aquatic ecosystems.

Contribution to Conservation:

By keeping swordtails in captivity and participating in responsible breeding practices, aquarists contribute to the conservation of freshwater fish species and their natural habitats.

Supporting conservation efforts helps raise awareness and promote sustainability within the aquarium hobby.

Overall, swordtail fish exhibit a combination of striking physical features, active behavior, and ease of care that make them popular and beloved among aquarium enthusiasts of all levels. With proper husbandry and attention to their needs, swordtails can thrive and bring joy to their keepers for years to come.

Selecting the Right Swordtail for You

Choosing the perfect swordtail fish (Xiphophorus hellerii) for your aquarium involves considering several factors to ensure compatibility and satisfaction. Here's a guide to help you make the right selection:

Tank Size and Space:

Assess the size of your aquarium and its inhabitants to determine how many swordtails you can accommodate.

Ensure that the tank has enough space for swordtails to swim and explore comfortably, as they are active fish.

Color and Appearance:

Consider the coloration and pattern preferences you have for your swordtail fish.

Choose varieties that appeal to you aesthetically and complement the overall look of your aquarium.

Gender and Social Dynamics:

Decide whether you want to keep male, female, or a mix of both swordtails in your tank.

Keep in mind that male swordtails often display more vibrant colors and have the distinctive sword-like extension on their tails, while females are typically larger and lack the sword.

Compatibility with Tankmates:

Research the compatibility of swordtails with other fish species in your aquarium.

Avoid keeping swordtails with aggressive or fin-nipping species that may harass or stress them.

Breeding Potential:

If you're interested in breeding swordtails, select healthy and robust specimens with desirable traits.

Consider setting up a separate breeding tank to facilitate successful breeding and raise fry.

Health and Condition:

Inspect the overall health and condition of the swordtail fish before purchasing them.

Look for signs of disease, such as lethargy, fin damage, abnormal behavior, or abnormal growths.

Source and Quality:

Purchase swordtail fish from reputable sources, such as trusted local fish stores or breeders.

Avoid buying fish from overcrowded or poorly maintained tanks, as they may carry diseases or parasites.

Personal Preferences and Goals:

Ultimately, choose swordtail fish that align with your personal preferences, goals, and aspirations as an aquarist.

Consider factors such as breeding potential, coloration, behavior, and long-term care requirements.

By carefully considering these factors and conducting thorough research, you can select the right swordtail fish for your aquarium that will thrive and bring joy to your fishkeeping journey. Remember to provide proper care, attention, and a suitable environment to ensure the health and well-being of your swordtails.

CHAPTER 3: Setting Up the Ideal Swordtail Aquarium

Creating a suitable environment for swordtail fish (Xiphophorus hellerii) involves careful consideration of their habitat preferences and specific needs. Here's a step-by-step guide to setting up the ideal swordtail aquarium:

Selecting the Tank:

Choose an aquarium with a sufficient size to accommodate the number of swordtails you plan to keep.

A tank size of at least 20 gallons (75 liters) is recommended for a small group of swordtails.

Aquascape and Decor:

Provide ample hiding spots and cover for swordtails by incorporating live or artificial plants, driftwood, rocks, and caves.

Create a naturalistic aquascape with a mix of open swimming areas and dense vegetation to mimic their native habitat.

Substrate and Lighting:

Use a fine gravel or sand substrate to allow foraging and rooting behavior.

Provide moderate lighting, supplemented with floating plants or shaded areas to reduce glare and mimic dappled sunlight.

Filtration and Water Quality:

Install a reliable filtration system to maintain water quality and circulation.

Swordtails prefer clean, well-oxygenated water with a pH range between 7.0 to 8.0 and temperatures between 72°F to 82°F (22°C to 28°C).

Heating and Thermoregulation:

Use a submersible heater to maintain stable water temperatures within the recommended range.

Provide areas of varying temperatures within the tank, such as cooler shaded spots and warmer areas near the heater, to allow swordtails to thermoregulate.

Water Parameters and Maintenance:

Regularly monitor water parameters, including ammonia, nitrite, nitrate, pH, and temperature, using reliable test kits.

Perform weekly partial water changes of 25% to 30% to remove accumulated waste and replenish essential nutrients.

Aquarium Mates and Compatibility:

Choose tank mates that are peaceful and compatible with swordtails, such as other community fish species that inhabit similar water conditions.

Avoid aggressive or fin-nipping species that may harass or stress swordtails.

Feeding and Nutrition:

Offer a varied diet consisting of high-quality commercial flakes, pellets, and frozen or live foods suitable for omnivorous fish.

Supplement their diet with fresh vegetables like blanched spinach or zucchini to provide essential nutrients and promote natural foraging behavior.

Observation and Maintenance:

Regularly observe swordtail behavior and health to detect any signs of stress, disease, or aggression.

Perform routine aquarium maintenance, including water changes, substrate vacuuming, and equipment checks, to ensure a healthy and thriving environment.

By following these guidelines and providing a well-planned and maintained aquarium, you can create the ideal habitat for swordtail fish to flourish and display their natural behaviors in captivity. Remember to consider their specific needs and preferences to promote their health, happiness, and longevity in the aquarium.

Tank Size and Setup for Swordtail Fish

Choosing the right tank size and setup is crucial for creating a suitable environment for swordtail fish (Xiphophorus hellerii). Here's how to determine the ideal tank size and set up for swordtails:

Tank Size:

A minimum tank size of 20 gallons (75 liters) is recommended for a small group of swordtail fish.

Larger tanks, such as 30 gallons (113 liters) or more, provide more swimming space and allow for a greater number of swordtails and tankmates.

Aquarium Setup:

Select a sturdy aquarium stand or cabinet that can support the weight of the tank and equipment.

Place the tank in a location away from direct sunlight and temperature fluctuations to prevent algae growth and temperature spikes.

Level the tank using shims or adjustable feet to ensure stability and even weight distribution.

Substrate:

Use a fine gravel or sand substrate, preferably dark in color, to mimic the natural environment of swordtail fish.

- Avoid sharp or abrasive substrates that may injure the fish or damage their delicate fins.

Filtration:

Install a reliable filtration system capable of handling the tank's volume and bioload.

Choose a combination of mechanical, biological, and chemical filtration to maintain optimal water quality and clarity.

Consider using a canister filter or a powerful hang-on-back filter for effective filtration and water circulation.

Heating:

Use a submersible aquarium heater to maintain stable water temperatures within the range of 72°F to 82°F (22°C to 28°C).

Position the heater near the filter outlet to ensure even heat distribution throughout the tank.

Lighting:
Provide moderate lighting using full-spectrum LED or fluorescent lights to promote plant growth and enhance the colors of swordtail fish. Use a timer to regulate the photoperiod and simulate natural day-night cycles for the fish and plants.

Decoration and Aquascape:
Create a naturalistic aquascape with a mix of live or artificial plants, driftwood, rocks, and caves.
Incorporate dense vegetation, such as java fern, amazon sword, and hornwort, to provide hiding spots and cover for swordtails.

Arrange the decor to create visual interest and break up sightlines, reducing stress and aggression among tankmates.

Water Parameters:

Maintain stable water parameters within the recommended range for swordtail fish: pH 7.0 to 8.0, hardness 10-25 dGH, and temperature 72°F to 82°F (22°C to 28°C).

Regularly test water parameters using reliable test kits and adjust as necessary with partial water changes and water conditioner.

By following these guidelines and providing a well-planned tank setup, you can create an ideal habitat for swordtail fish to thrive and display their natural behaviors in captivity. Remember to regularly monitor water quality, perform routine maintenance, and observe the fish for signs of health or behavior issues to ensure their well-being and longevity in the aquarium.

Water Parameters and Filtration for Swordtail Fish

Maintaining optimal water parameters and filtration are essential for the health and well-being of swordtail fish (Xiphophorus hellerii). Here's how to ensure proper water parameters and filtration in your aquarium:

Water Parameters:

pH: Swordtails prefer slightly alkaline water with a pH range between 7.0 to 8.0.

Temperature: Maintain stable water temperatures between 72°F to 82°F (22°C to 28°C) using a submersible aquarium heater.

Hardness: Swordtails thrive in moderately hard water with a hardness level (GH) of 10-25 dGH.
Ammonia, Nitrite, and Nitrate: Keep ammonia and nitrite levels at zero ppm (parts per million) and nitrate levels below 40 ppm through regular water testing and partial water changes.

Filtration:

Mechanical Filtration: Use filter media such as filter floss or sponges to trap debris and particulate matter from the water column.

Biological Filtration: Provide ample surface area for beneficial bacteria to colonize and break down harmful ammonia and nitrite into less toxic nitrate. Ceramic rings, bio-balls, or porous sponge filters are ideal for biological filtration.

Chemical Filtration: Consider using activated carbon or zeolite media to remove dissolved organic compounds, odors, and impurities from the water.

Filtration Capacity: Choose a filtration system rated for at least 4-6 times the volume of your aquarium to ensure adequate water turnover and filtration efficiency.

Maintenance: Clean or replace filter media regularly to prevent clogging and maintain optimal filtration performance. Rinse mechanical media in aquarium water to preserve beneficial bacteria.

Water Circulation:

Ensure adequate water circulation throughout the aquarium to prevent stagnant areas and promote oxygenation.

Position the filter outlet and aquarium air stones strategically to create gentle water movement and surface agitation.

Aquarium Placement:

Place the aquarium in a stable location away from direct sunlight and temperature fluctuations to prevent algae growth and temperature spikes.

 Use a sturdy aquarium stand or cabinet that can support the weight of the tank and equipment.

Monitoring and Maintenance:

Regularly test water parameters using reliable test kits to monitor ammonia, nitrite, nitrate, pH, and temperature.

Perform weekly partial water changes of 25% to 30% to remove accumulated waste and replenish essential nutrients.

Clean filter intake tubes, impellers, and hoses periodically to prevent blockages and ensure efficient water flow.

By maintaining optimal water parameters and filtration, you can create a healthy and stable environment for swordtail fish to thrive and display their natural behaviors in the aquarium. Regular monitoring and maintenance are key to preventing water quality issues and promoting the well-being of your fish.

Substrate, Decorations, and Plants for Swordtail Fish Aquarium

Creating an enriched and naturalistic environment for swordtail fish (Xiphophorus hellerii) involves carefully selecting substrate, decorations, and live plants. Here's how to set up the substrate, decorations, and plants in your swordtail aquarium:

Substrate:

Choose a fine gravel or sand substrate, preferably dark in color, to mimic the natural environment of swordtail fish.

Avoid sharp or abrasive substrates that may injure the fish or damage their delicate fins.

Ensure a depth of at least 2-3 inches (5-7.5 cm) to allow for rooting and foraging behavior.

Decorations:

Provide ample hiding spots and cover for swordtails by incorporating a variety of decorations such as driftwood, rocks, caves, and PVC pipes.

Arrange the decorations to create visual interest and break up sightlines, reducing stress and aggression among tankmates.

Use natural materials like aquarium-safe driftwood and rocks to create a more authentic and aesthetically pleasing environment.

Live Plants:

Incorporate live plants to provide shelter, oxygenation, and natural filtration in the aquarium.

Choose hardy and fast-growing plant species that can tolerate the water conditions and grazing behavior of swordtail fish, such as:

Java fern (Microsorum pteropus)

Anubias (Anubias spp.)

Java moss (Taxiphyllum barbieri)

Hornwort (Ceratophyllum demersum)

Amazon sword (Echinodorus spp.)

Plant the live plants directly into the substrate or attach them to driftwood or rocks using aquarium-safe glue or thread.

Maintain proper lighting and nutrient levels to promote healthy plant growth and prevent algae overgrowth.

Artificial Decorations:

If live plants are not feasible, consider using high-quality artificial plants made from aquarium-safe materials.

Choose artificial decorations that closely resemble natural plants and provide similar benefits in terms of shelter and visual appeal.

Arrange artificial decorations strategically to create hiding spots and break up sightlines, enhancing the overall aesthetics of the aquarium.

Aquascaping Tips:

Create a naturalistic aquascape by combining different substrate textures, decorations, and plants to mimic the diverse habitats of swordtail fish in the wild.

Use a focal point, such as a large rock or driftwood centerpiece, to anchor the aquascape and create a focal point of interest.

Experiment with different layouts and arrangements to find the most visually appealing and functional design for your swordtail aquarium.

By incorporating suitable substrate, decorations, and live plants, you can create a stimulating and enriched environment for swordtail fish to thrive and display their natural behaviors in captivity. Regular maintenance and care are essential to ensure the health and well-being of both the fish and the aquatic environment.

CHAPTER 4: Swordtail Fish Care Guide

Swordtail fish (Xiphophorus hellerii) are relatively easy to care for, making them suitable for both beginner and experienced aquarium enthusiasts. Here's a comprehensive care guide to help you provide the best possible care for your swordtail fish:

Tank Setup:

Tank Size: Provide a spacious aquarium with a minimum size of 20 gallons (75 liters) for a small group of swordtail fish.

Substrate: Use fine gravel or sand substrate, preferably dark in color, to mimic the natural environment of swordtails.

Filtration: Install a reliable filtration system capable of handling the tank's bioload, with mechanical, biological, and chemical filtration components.

Heating: Maintain stable water temperatures between 72°F to 82°F (22°C to 28°C) using a submersible aquarium heater.

Lighting: Provide moderate lighting using full-spectrum LED or fluorescent lights to promote plant growth and enhance fish colors.

Decoration: Create a naturalistic aquascape with live or artificial plants, driftwood, rocks, and caves to provide hiding spots and cover for swordtails.

Water Parameters: Maintain stable water parameters within the following ranges: pH 7.0 to 8.0, hardness (GH) 10-25 dGH, and temperature 72°F to 82°F (22°C to 28°C).

Feeding:

Offer a varied diet consisting of high-quality commercial flakes, pellets, and frozen or live foods suitable for omnivorous fish.

Supplement their diet with fresh vegetables like blanched spinach, zucchini, or peas to provide essential nutrients and promote natural foraging behavior.

Feed small amounts of food 2-3 times a day, ensuring that all fish have an opportunity to eat without overfeeding.

Tankmates:

Choose peaceful and compatible tank mates for swordtail fish, such as other community fish species that inhabit similar water conditions.

Avoid aggressive or fin-nipping species that may harass or stress swordtails.

Consider keeping swordtails in groups of their own species or with other livebearers like mollies, platies, or guppies.

Breeding:

Swordtails are prolific breeders, with females capable of producing multiple batches of fry throughout their lifespan.

Provide suitable breeding conditions, such as ample hiding places for fry and a separate breeding tank to protect them from predators.

Monitor water parameters and maintain optimal tank conditions to encourage successful breeding and fry survival.

Health and Maintenance:

Regularly monitor water parameters using reliable test kits and perform weekly partial water changes of 25% to 30% to maintain water quality.

Conduct routine maintenance tasks such as substrate vacuuming, filter cleaning, and equipment checks to ensure a healthy and thriving aquarium.

Keep an eye out for signs of disease or illness, such as lethargy, loss of appetite, abnormal behavior, or visible symptoms, and address any issues promptly with appropriate treatment.

By following this care guide and providing a well-maintained environment, you can ensure the health, happiness, and longevity of your swordtail fish in the aquarium. Remember to observe them regularly and adjust care practices as needed to meet their specific needs and preferences.

Feeding Swordtails

Feeding swordtail fish (Xiphophorus hellerii) a balanced and varied diet is essential for their health and well-being. Here are some tips for feeding swordtails:

Commercial Flakes or Pellets:
Offer high-quality commercial flakes or pellets formulated specifically for freshwater fish.
Look for products that contain a balanced mix of protein, vitamins, and minerals to meet the nutritional needs of swordtails.

Feed small amounts of flakes or pellets 2-3 times a day, only giving them what they can consume within a few minutes to prevent overfeeding.

Frozen or Live Foods:
Supplement their diet with frozen or live foods such as bloodworms, brine shrimp, daphnia, or tubifex worms.
These foods provide essential protein and variety, mimicking their natural diet in the wild.
Offer frozen or live foods as occasional treats 2-3 times a week to provide enrichment and stimulate natural feeding behaviors.

Vegetables:
Include fresh vegetables in their diet to provide essential nutrients and promote digestion.
Blanched vegetables such as spinach, zucchini, cucumber, or peas are suitable options for swordtails.

Cut the vegetables into small pieces and blanch them by briefly boiling or steaming until softened before offering them to the fish.

Algae-Based Foods:

Swordtails may also graze on algae growing in the aquarium, but it should not be their primary source of nutrition.

Supplement their diet with algae-based sinking pellets or wafers to ensure they receive enough fiber and plant matter.

Variety is Key:

Offer a diverse diet consisting of a combination of flakes, pellets, frozen or live foods, and vegetables to ensure nutritional balance.

Rotate their diet regularly to provide variety and prevent nutritional deficiencies or boredom.

Observation and Adjustments:

Observe swordtail fish during feeding times to ensure that all individuals are eating and competing for food.

Adjust feeding amounts and frequency based on the fish's appetite and behavior, being mindful not to overfeed.

Water Quality and Maintenance:

Remove any uneaten food from the tank after feeding to prevent it from decomposing and affecting water quality.

Monitor water parameters regularly using reliable test kits and perform routine water changes to maintain optimal water quality and a healthy environment for swordtails.

By providing a balanced and varied diet, you can ensure that swordtail fish receive the essential nutrients they need to thrive and display their vibrant colors and natural behaviors in the aquarium.

Water Maintenance and Testing for Swordtail Fish

Maintaining optimal water quality is crucial for the health and well-being of swordtail fish (Xiphophorus hellerii). Here's a guide to water maintenance and testing to ensure a healthy aquarium environment:

Water Testing:
Regularly test water parameters using reliable test kits to monitor key parameters such as:
Ammonia (NH_3/NH_4^+)
Nitrite (NO_2^-)
Nitrate (NO_3^-)
pH
Temperature
General hardness (GH)

Carbonate hardness (KH)

Test water parameters weekly or bi-weekly, especially during the initial setup of the aquarium and after any significant changes or additions.

Water Changes:

Perform weekly partial water changes of 25% to 30% to remove accumulated waste, uneaten food, and excess nutrients.

Use a siphon or gravel vacuum to clean the substrate and remove debris from the bottom of the tank.

Treat tap water with a water conditioner to remove chlorine, chloramine, and heavy metals before adding it to the aquarium.

Filter Maintenance:

Clean or replace filter media regularly to prevent clogging and maintain optimal filtration performance.

Rinse mechanical media in aquarium water to preserve beneficial bacteria and avoid disrupting the biological filtration process.

Check filter intake tubes, impellers, and hoses for debris and clean them as needed to ensure efficient water flow.

Algae Control:

Control algae growth by maintaining stable water parameters, proper lighting, and nutrient levels.

Remove excess nutrients through regular water changes and by avoiding overfeeding.

Use algae scrapers or algae-eating fish like otocinclus catfish or Siamese algae eaters to help keep algae in check.

Plant Care:

If you have live plants in the aquarium, trim them regularly to remove dead or decaying plant matter and promote healthy growth.

Monitor plant health and address any nutrient deficiencies or algae issues promptly to prevent them from spreading.

Water Conditioners and Additives:

Use water conditioners to remove chlorine, chloramine, and other harmful substances from tap water before adding it to the aquarium.

Consider using additional additives such as aquarium salt or products to enhance water quality, promote fish health, and reduce stress.

Monitoring and Observation:

Keep a close eye on swordtail fish behavior, appearance, and appetite to detect any signs of stress, disease, or water quality issues.

Address any abnormalities or concerns promptly by testing water parameters and taking appropriate corrective actions.

By implementing a regular water maintenance routine and testing water parameters, you can create a stable and healthy environment for swordtail fish to thrive and display their natural behaviors in the aquarium. Consistent monitoring and adjustments will help ensure the long-term health and well-being of your fish.

Handling Health Issues

Handling health issues promptly is essential for maintaining the well-being of swordtail fish (Xiphophorus hellerii) in the aquarium. Here's a guide to handling common health issues:

Identification:

Monitor swordtail fish closely for any signs of illness or abnormal behavior, such as:

Lethargy or lack of activity

Loss of appetite or refusal to eat

Erratic swimming behavior

Gasping at the water surface

Changes in coloration or appearance

Visible signs of disease such as white spots (ich), fin rot, or fungal growths.

Isolation:

If you suspect that a fish is sick, consider isolating it in a separate quarantine tank to prevent the spread of disease to other tankmates.

Quarantine tanks should be equipped with appropriate filtration, heating, and aeration, and maintained with regular water changes.

Water Quality:

Test water parameters using reliable test kits to ensure that water quality is within acceptable ranges.

Address any water quality issues promptly through water changes, filtration maintenance, and proper feeding practices.

Treatment Options:

Consult a veterinarian or experienced aquarist for accurate diagnosis and treatment recommendations based on the specific symptoms and underlying causes.

Treatments may include medications, antibiotics, antiparasitic agents, or antifungal treatments, administered according to the manufacturer's instructions and dosage recommendations.

Follow proper quarantine procedures when administering medications to avoid affecting other tank inhabitants or disrupting the biological filtration process.

Quarantine Protocol:

Quarantine newly acquired fish for a minimum of two weeks before introducing them to the main aquarium to prevent the introduction of diseases.

Observe quarantined fish closely for any signs of illness and treat as needed before transferring them to the main tank.

Preventive Measures:

Practice good aquarium husbandry and hygiene to minimize the risk of disease outbreaks, including regular water changes, filter maintenance, and substrate vacuuming.

Avoid overstocking the aquarium and maintain appropriate stocking levels to reduce stress and competition among fish.

Quarantine and treat new fish before introducing them to the main tank to prevent the spread of diseases.

Professional Assistance:

If you're unsure about the diagnosis or treatment of a health issue, seek advice from a qualified veterinarian with experience in fish health and aquatic medicine.

Online forums and community groups can also provide valuable insights and support from experienced hobbyists who have dealt with similar health issues.

By promptly identifying and addressing health issues, providing appropriate treatment, and implementing preventive measures, you can help ensure the health and well-being of swordtail fish in the aquarium. Regular monitoring and attentive care will contribute to a thriving and vibrant aquatic environment for your fish to enjoy.

CHAPTER 5: Swordtail Breeding

Breeding swordtail fish (Xiphophorus hellerii) can be a rewarding experience for aquarium hobbyists. Here's a guide to swordtail breeding:

Gender Identification:
Swordtails are livebearers, meaning females give birth to live young rather than laying eggs. Male swordtails can be distinguished by their gonopodium, a modified anal fin used for mating, which is more elongated and pointed compared to the female's rounded anal fin.

Breeding Setup:
Set up a separate breeding tank to facilitate successful breeding and protect fry from predators.
Use a smaller tank (10-20 gallons) with a sponge filter or gentle filtration system to prevent fry from being sucked into the filter.

Conditioning:

Condition breeding pairs with a high-quality diet rich in protein and vitamins to ensure optimal health and reproductive readiness.

Provide ample hiding spots and plants in the breeding tank to create a conducive environment for breeding and fry survival.

Introducing the Pair:

Introduce a compatible male and female swordtail to the breeding tank, ensuring that both are mature and healthy.

Monitor their behavior closely for signs of courtship, such as chasing and displaying vibrant colors.

Mating and Gestation:

During mating, the male swordtail will use his gonopodium to fertilize the female's eggs internally.

Female swordtails can store sperm from multiple matings, allowing them to produce multiple batches of fry without repeated mating. Gestation typically lasts 4-6 weeks, depending on factors such as water temperature and the individual female's health.

Fry Care:

Once the female gives birth, move her back to the main tank to prevent her from consuming the fry.

Fries are typically born fully formed and able to swim immediately.

Provide ample hiding spots and floating plants in the breeding tank to protect fry from being eaten by adult fish.

Feeding Fry:

Feed fry small, frequent meals of powdered or liquid fry food, infusoria, or newly hatched brine shrimp.

Gradually introduce finely crushed flakes or pellets as the fry grow larger.

Growth and Development:

Monitor fry growth and development closely, ensuring they have access to proper nutrition and a clean environment.

Perform regular water changes and maintain stable water parameters to promote healthy growth and minimize stress.

Selective Breeding:

Selectively breed swordtails with desirable traits such as vibrant colors, unique patterns, or specific fin shapes.

Keep detailed records of breeding pairs and offspring to track lineage and genetic traits.

Breeding swordtail fish can be a fascinating and enjoyable aspect of the aquarium hobby. With proper care, attention to detail, and a well-planned breeding setup, you can successfully raise healthy swordtail fry and contribute to the diversity of this popular freshwater fish species.

Breeding Basics

Breeding swordtail fish (Xiphophorus hellerii) involves understanding the basics of their reproductive behavior, setup requirements, and care for fry. Here's a breakdown of the breeding basics for swordtails:

Reproductive Behavior:
Swordtails are livebearers, meaning females give birth to live young rather than laying eggs.

Males have a gonopodium, a modified anal fin used for mating, which they use to fertilize the female's eggs internally.

Courtship behavior includes chasing, displaying vibrant colors, and occasional flaring of fins by males to attract females.

Breeding Setup:

Set up a separate breeding tank to facilitate successful breeding and protect fry from predators.

Use a smaller tank (10-20 gallons) with gentle filtration and ample hiding spots such as plants or caves.

Maintain stable water parameters with a temperature of 72°F to 82°F (22°C to 28°C) and pH of 7.0 to 8.0.

Selecting Breeding Pairs:

Choose healthy, mature male and female swordtails for breeding, ensuring they are free from diseases or deformities.

Select breeding pairs based on desired traits such as coloration, fin shape, or pattern.

Introducing the Pair:

Introduce the breeding pair to the breeding tank and monitor their behavior closely.

Males will typically pursue females, displaying courtship behavior such as flashing colors and fin displays.

Mating and Gestation:

During mating, the male will use his gonopodium to fertilize the female's eggs internally.

Female swordtails can store sperm from multiple matings, allowing them to produce multiple batches of fry without repeated mating.

Gestation typically lasts 4-6 weeks, during which the female's abdomen will gradually swell as the fry develop.

Fry Care:
Once the female gives birth, remove her from the breeding tank to prevent her from consuming the fry.
Provide hiding spots and floating plants in the breeding tank to protect fry from being eaten by adult fish.
Feed fry small, frequent meals of powdered or liquid fry food, infusoria, or newly hatched brine shrimp.

Growth and Development:
Monitor fry growth and development closely, ensuring they have access to proper nutrition and a clean environment.

Perform regular water changes and maintain stable water parameters to promote healthy growth and minimize stress.

Separate large fry from smaller ones if necessary to prevent cannibalism and ensure adequate food availability for all.

Breeding swordtail fish can be a fascinating and rewarding experience for aquarium hobbyists. By understanding the basics of swordtail reproduction and providing proper care for breeding pairs and fry, you can successfully raise healthy generations of these colorful freshwater fish in your aquarium.

Breeding Tank Setup

Setting up a breeding tank for swordtail fish requires careful consideration of their breeding behavior and specific needs. Here's how to set up a breeding tank for swordtails:

Tank Size:

Choose a tank with a capacity of at least 10-20 gallons (38-75 liters) to provide enough space for breeding pairs and fry.

Larger tanks offer more stability and room for growth, especially if you plan to raise multiple batches of fry simultaneously.

Filtration:

Use a gentle filtration system such as a sponge filter or a low-flow hang-on-back filter to avoid sucking up fry.

Ensure that the filter provides adequate biological filtration to maintain water quality in the breeding tank.

Substrate:

Use a fine gravel or sand substrate to mimic the natural environment of swordtail fish.

Avoid sharp or abrasive substrates that may damage their delicate fins.

Decorations:

Provide ample hiding spots and cover for breeding pairs and fry by incorporating live or artificial plants, driftwood, rocks, and caves.

Arrange decorations to create visual barriers and break up sightlines, reducing stress and aggression among tank inhabitants.

Use floating plants such as java moss or water lettuce to provide additional cover and protect fry from being eaten by adult fish.

Lighting:

Use moderate lighting with full-spectrum LED or fluorescent lights to promote plant growth and simulate natural daylight.

Avoid excessively bright lighting, as it may cause stress or algae overgrowth in the breeding tank.

Water Parameters:

Maintain stable water parameters within the following ranges:

Temperature: 72°F to 82°F (22°C to 28°C)

pH: 7.0 to 8.0

Hardness (GH): 10-25 dGH

Ammonia, Nitrite, Nitrate: Keep levels as close to zero as possible through regular water changes and proper filtration.

Breeding Setup:

Introduce a compatible breeding pair of swordtail fish to the tank, ensuring that both are healthy and mature.

Provide ample hiding spots and cover for the female to give birth and for fry to seek refuge after birth.

Monitor breeding behavior closely, including courtship displays, mating attempts, and signs of pregnancy in the female.

Maintenance:

Perform regular water changes of 25% to 30% weekly to maintain water quality and remove accumulated waste.

Clean or replace filter media as needed to ensure optimal filtration performance and water clarity.

Monitor water parameters regularly using reliable test kits and adjust as necessary to maintain optimal conditions for breeding and fry development.

By setting up a dedicated breeding tank with appropriate equipment, decorations, and water parameters, you can provide a safe and conducive environment for swordtail fish to breed and raise their fry successfully. Regular maintenance and observation will help ensure the health and well-being of breeding pairs and their offspring in the aquarium.

Caring for Swordtail Fry

Caring for swordtail fry (baby swordtails) requires attention to their specific needs and providing a safe environment for their growth and development. Here's a guide to caring for swordtail fry:

Feeding:

Feed swordtail fry small, frequent meals of specialized fry food, powdered flakes, or liquid fry food.

Offer newly hatched brine shrimp or infusoria as a nutritious and natural food source for fry.

Gradually introduce finely crushed flakes or pellets as the fry grow larger.

Feed multiple times a day to ensure they receive enough nutrition for growth and development.

Water Quality:

Maintain stable water parameters within the following ranges:

Temperature: 72°F to 82°F (22°C to 28°C)

pH: 7.0 to 8.0

Hardness (GH): 10-25 dGH

Ammonia, Nitrite, Nitrate: Keep levels as close to zero as possible through regular water changes and proper filtration.

Perform small, frequent water changes to prevention ammonia buildup and maintain optimal water quality for fry.

Tank Setup:

Provide ample hiding spots and cover for fry by incorporating live or artificial plants, floating plants, and fine-leaved vegetation.

Use a sponge filter or a low-flow filtration system to avoid sucking up fry.

Ensure that the tank has a secure lid or cover to prevent fry from jumping out.

Tankmates:

Keep swordtail fry separate from adult fish, as they may be eaten or harassed by larger tankmates.

Avoid housing aggressive or predatory fish species with swordtail fry to minimize stress and predation risk.

Growth and Development:

Monitor fry growth and development closely, ensuring they have access to proper nutrition and a clean environment.

Separate large fry from smaller ones if necessary to prevent cannibalism and ensure adequate food availability for all.

Provide enrichment and stimulation by varying their diet, introducing different foods, and occasionally changing tank decor.

Water Changes:

Perform regular water changes of 25% to 30% weekly to maintain water quality and remove accumulated waste.

Use a siphon or gravel vacuum to clean the substrate and remove debris from the bottom of the tank.

Observation and Care:

Monitor fry behavior, appearance, and appetite daily to detect any signs of illness or stress.

Address any abnormalities or concerns promptly by adjusting water parameters, providing proper nutrition, and taking appropriate corrective actions.

By providing a well-maintained environment, proper nutrition, and attentive care, you can ensure the healthy growth and development of swordtail fry in the aquarium. With time and dedication, you can enjoy watching them grow into colorful and vibrant adult fish.

CHAPTER 6: Compatibility and Tankmates

When selecting tankmates for swordtail fish (Xiphophorus hellerii), it's essential to choose species that are compatible in terms of temperament, size, and water parameters. Here's a guide to compatible tankmates for swordtails:

Peaceful Community Fish:

Swordtails are generally peaceful and can coexist with a variety of other community fish species.

Good tankmates include:

Mollies (Poecilia spp.)

Platies (Xiphophorus spp.)

Guppies (Poecilia reticulata)

Tetras (Neon tetras, Cardinal tetras, etc.)

Rasboras (Harlequin rasboras, Chili rasboras, etc.)

Corydoras catfish

Dwarf gouramis (Trichogaster lalius)

Rainbowfish (Melanotaeniidae spp.)

Avoid Aggressive or Predatory Species:

Avoid keeping swordtails with aggressive or predatory species that may harass or prey on them.

Species to avoid include:

Cichlids (except for peaceful species like dwarf cichlids)

Larger predatory fish such as angelfish, Oscars, or larger catfish species

Fin-nipping species such as tiger barbs or some species of danios

Similar Water Parameter Requirements:

Choose tank mates that have similar water parameter requirements to swordtails to minimize stress and health issues.

Match temperature, pH, and hardness preferences to create a harmonious environment for all tank inhabitants.

Consider Tank Size and Space:
Ensure that the aquarium is large enough to accommodate all tankmates comfortably and provide ample swimming space.
Avoid overstocking the tank to prevent aggression, competition for resources, and water quality issues.

Compatibility with Plants:
Consider the compatibility of tankmates with live plants if you have a planted aquarium.
Avoid species known for uprooting or damaging plants, such as some species of cichlids or larger catfish.

Observing Compatibility:

Introduce new tankmates gradually and monitor their interactions closely for any signs of aggression or stress.

Be prepared to remove or rehome incompatible tank mates if conflicts arise.

By selecting compatible tank mates and providing a harmonious environment, you can create a peaceful and thriving community aquarium with swordtail fish as the centerpiece. Regular observation and maintenance will help ensure the well-being of all tank inhabitants.

Suitable Tankmates for Swordtails

Swordtails (Xiphophorus hellerii) are generally peaceful fish that can coexist with a variety of other community fish species. Here's a list of suitable tankmates for swordtails:

Mollies (Poecilia spp.):

Mollies are closely related to swordtails and share similar care requirements, making them excellent tankmates.

They come in various colors and patterns, adding diversity to the aquarium.

Platies (Xiphophorus spp.):

Platies are another livebearer species closely related to swordtails and are compatible in terms of temperament and care.

They are available in a wide range of colors and are easy to care for, making them popular choices for community tanks.

Guppies (Poecilia reticulata):

Guppies are colorful and active fish that thrive in community aquariums.

They share similar water parameter requirements with swordtails and can coexist peacefully in the same tank.

Tetras (e.g., Neon Tetras, Cardinal Tetras):

Small tetra species like neon tetras and cardinal tetras make excellent tankmates for swordtails.

They are peaceful, schooling fish that add movement and color to the aquarium.

Rasboras (e.g., Harlequin Rasboras, Chili Rasboras):

Rasboras are another group of peaceful, schooling fish that are compatible with swordtails.

They come in various sizes and colors, adding visual interest to the tank.

Corydoras Catfish:

Corydoras catfish are bottom-dwelling fish that help keep the tank clean by scavenging for leftover food.

They are peaceful and can coexist with swordtails without any issues.

Dwarf Gouramis (Trichogaster lalius):

Dwarf gouramis are colorful and relatively peaceful labyrinth fish that can inhabit the upper levels of the aquarium.

They add visual interest and diversity to the tank without causing problems for swordtails.

Rainbowfish (Melanotaeniidae spp.):

Rainbowfish are active and colorful fish that thrive in community setups.

They come in various sizes, so be sure to choose species that are compatible with the size of your tank.

When selecting tankmates for swordtails, it's essential to consider factors such as tank size, compatibility in terms of temperament and water parameters, and the potential for aggression or fin-nipping behavior. By choosing suitable tankmates, you can create a harmonious community aquarium where swordtails can thrive alongside other fish species.

Potential Compatibility Issues

While swordtails (Xiphophorus hellerii) are generally peaceful fish, there are some potential compatibility issues to consider when selecting tankmates:

Aggressive Species: Avoid keeping swordtails with aggressive or territorial fish species that may harass or intimidate them. Aggressive species can cause stress and physical harm to swordtails, leading to health issues and reduced well-being.

Fin-Nipping Fish: Some fish species, such as tiger barbs or certain types of tetras, are known for their fin-nipping behavior. This can be particularly problematic for swordtails, as their long, flowing fins make them vulnerable to injury. Avoid keeping swordtails with fin-nipping species to prevent damage to their fins and reduce stress.

Predatory Fish: Larger predatory fish species, such as angelfish, Oscars, or larger cichlids, may view swordtails as potential prey. Keeping swordtails with predatory fish can result in aggression, predation, and stress for the swordtails. It's best to avoid mixing swordtails with predatory species to ensure their safety.

Size Discrepancies: Avoid pairing swordtails with fish species that are significantly larger or smaller than them. Larger fish may view swordtails as potential prey, while smaller fish may be intimidated or harassed by swordtails. Choose tank mates that are similar in size to swordtails to promote peaceful coexistence.

Incompatible Water Parameters: Swordtails thrive in freshwater with slightly alkaline pH and moderate hardness. Avoid keeping them with fish species that have significantly different water parameter requirements, as this can lead to stress and health issues for both swordtails and their tankmates. Choose tankmates with similar water parameter preferences to ensure compatibility.

Overcrowding: Overcrowding can lead to aggression, competition for resources, and stress among tank inhabitants, including swordtails. Ensure that the aquarium is adequately sized and stocked to accommodate all fish species comfortably. Avoid overstocking the tank to prevent compatibility issues and promote a harmonious community environment.

By carefully selecting compatible tank mates and ensuring that the aquarium is properly maintained, you can create a peaceful and thriving community aquarium for swordtails and other fish species to coexist harmoniously. Regular observation and monitoring will help you identify and address any potential compatibility issues before they escalate.

Creating a Harmonious Community Tank

Creating a harmonious community tank involves carefully selecting compatible fish species, providing appropriate habitat and resources, and maintaining optimal water quality. Here's a step-by-step guide to creating a harmonious community tank:

Research Fish Species:

Research the compatibility, temperament, and care requirements of potential fish species before adding them to the tank.

Consider factors such as size, activity level, feeding habits, and water parameter preferences.

Choose Compatible Species:
Select fish species that are known to coexist peacefully and have similar temperaments and water parameter requirements.
Avoid mixing aggressive, fin-nipping, or predatory species with peaceful community fish.

Plan Tank Setup:
Design the aquarium layout to provide ample swimming space, hiding spots, and territories for each fish species.
Use plants, rocks, driftwood, and other decorations to create natural barriers and visual interest in the tank.

Consider Tank Size:

Choose an appropriately sized tank based on the number and size of fish species you plan to keep.

Larger tanks provide more space for fish to establish territories and reduce aggression.

Introduce Fish Gradually:

Add fish to the tank gradually, starting with the least aggressive or territorial species first.

Monitor fish behavior closely after each addition to ensure compatibility and prevent aggression.

Monitor Water Quality:

Test water parameters regularly using reliable test kits to monitor ammonia, nitrite, nitrate, pH, and temperature levels.

Perform regular water changes to maintain optimal water quality and prevent the buildup of toxins.

Provide Balanced Diet:

Offer a varied and balanced diet to meet the nutritional needs of all fish species in the tank.

Feed a combination of high-quality flakes, pellets, frozen or live foods, and fresh vegetables.

Observe and Adjust:

Observe fish behavior and interactions regularly to identify any signs of aggression, stress, or compatibility issues.

Be prepared to rehome or separate fish if conflicts arise or if certain species become incompatible over time.

Maintain Peaceful Environment:

Minimize stressors in the tank, such as sudden changes in water parameters, overcrowding, or aggressive tankmates.

Provide adequate filtration, aeration, and hiding spots to reduce stress and promote natural behavior.

Seek Advice When Needed:

Consult with experienced aquarists or seek advice from reputable sources if you encounter compatibility issues or have questions about specific fish species.

Online forums, local fish stores, and aquarium clubs can provide valuable insights and support.

By following these steps and maintaining a careful balance of fish species and environmental conditions, you can create a harmonious community tank where fish coexist peacefully and thrive in a natural and healthy environment.

CHAPTER 7: Tips for Success with Swordtails

To ensure success with swordtail fish (Xiphophorus hellerii) in your aquarium, consider the following tips:

Provide Adequate Space: Swordtails are active swimmers and require ample space to thrive. Choose a tank size of at least 20 gallons (75 liters) for a small group of swordtails to allow them plenty of room to swim and establish territories.

Maintain Stable Water Parameters: Swordtails prefer slightly alkaline water with a pH ranging from 7.0 to 8.0 and moderate hardness. Keep water temperature between 72°F to 82°F (22°C to 28°C) and perform regular water changes to maintain stable water quality.

Offer Varied Diet: Provide a balanced diet consisting of high-quality flakes, pellets, and occasional live or frozen foods such as bloodworms, brine shrimp, or daphnia. Include vegetable matter in their diet, such as blanched spinach or zucchini, to ensure they receive essential nutrients.

Create Hiding Spots: Include plenty of plants, driftwood, and decorations in the aquarium to provide hiding spots and cover for swordtails. They appreciate dense vegetation to retreat to when stressed or to rest.

Monitor Tank Mates: Choose tank mates that are compatible with swordtails in terms of temperament and size. Avoid keeping aggressive or fin-nipping species with swordtails to prevent stress and injury.

Maintain Good Filtration: Use a reliable filtration system to keep the water clean and well-oxygenated. Swordtails are sensitive to poor water quality, so ensure adequate filtration and perform regular maintenance to keep ammonia and nitrite levels low.

Avoid Overcrowding: Do not overcrowd the aquarium, as this can lead to stress, aggression, and competition for resources among fish. Follow stocking guidelines and provide enough space for swordtails to establish territories.

Observe Behavior: Monitor swordtail behavior regularly to detect any signs of illness, aggression, or stress. Address any issues promptly by identifying and addressing the underlying cause.

Provide Enrichment: Stimulate swordtails' natural behaviors by incorporating toys, floating plants, and areas to explore in the aquarium. This can help prevent boredom and promote overall well-being.

Breeding Considerations: If you plan to breed swordtails, provide a separate breeding tank with appropriate hiding spots and plants for fry to thrive. Monitor water parameters closely during breeding and provide proper care for fry as they grow.

By following these tips and providing appropriate care, you can create a thriving and healthy environment for swordtail fish in your aquarium.

Common Mistakes to Avoid

To ensure the health and well-being of swordtail fish (Xiphophorus hellerii) in your aquarium, here are some common mistakes to avoid:

Overstocking: Avoid overcrowding the aquarium with too many fish. Overstocking can lead to increased competition for resources, stress, aggression, and poor water quality.

Inadequate Tank Size: Providing insufficient space can stress swordtails and lead to aggressive behavior. Ensure your tank is appropriately sized for the number of fish you plan to keep.

Poor Water Quality: Neglecting water quality maintenance can result in stress, illness, and even death in swordtails. Perform regular water changes, monitor water parameters, and clean the tank and filter regularly.

Incompatible Tankmates: Choosing aggressive or incompatible tankmates can lead to stress, injuries, and conflicts among fish. Research and select tank mates that are peaceful and compatible with swordtails.

Inadequate Filtration: Insufficient filtration can lead to a buildup of harmful waste and toxins in the aquarium. Use a properly sized filter and perform regular maintenance to keep water clean and healthy for swordtails.

Poor Diet: Providing an inadequate diet can result in malnutrition and health problems in swordtails. Offer a varied diet consisting of high-quality flakes, pellets, and occasional live or frozen foods to meet their nutritional needs.

Lack of Hiding Places: Swordtails require hiding spots and cover to feel secure and reduce stress. Provide plenty of plants, caves, and decorations to create hiding places for them.

Ignoring Signs of Illness: Ignoring signs of illness or disease in swordtails can result in worsening health and spread to other fish. Monitor fish behavior and appearance regularly and address any signs of illness promptly.

Failure to Quarantine New Fish: Introducing new fish to the aquarium without quarantine can introduce diseases and parasites to the tank. Quarantine new fish for at least a few weeks before adding them to the main tank.

Skipping Water Changes: Neglecting water changes can lead to a buildup of toxins and pollutants in the aquarium. Perform regular water changes to maintain optimal water quality and promote the health of swordtails and other tank inhabitants.

By avoiding these common mistakes and providing proper care and attention, you can create a healthy and thriving environment for swordtail fish in your aquarium.

Enhancing the Well-Being of Your Swordtails

Enhancing the well-being of your swordtail fish (Xiphophorus hellerii) involves providing a stimulating and comfortable environment while attending to their physical and behavioral needs. Here are some tips to enhance the well-being of your swordtails:

Provide Adequate Space: Ensure your aquarium is appropriately sized for the number of swordtails you have. Swordtails are active swimmers and require ample space to move around. A tank size of at least 20 gallons (75 liters) is recommended for a small group of swordtails.

Maintain Water Quality: Regularly test and monitor water parameters such as temperature, pH, ammonia, nitrite, and nitrate levels. Perform regular water changes to keep water clean and healthy for your swordtails. Avoid overfeeding and remove uneaten food promptly to prevent water quality issues.

Offer Hiding Places: Provide plenty of hiding spots and cover in the aquarium using live or artificial plants, driftwood, caves, and other decorations. Hiding places help reduce stress and provide security for swordtails, especially when they feel threatened or stressed.

Provide Varied Diet: Offer a varied and nutritious diet to meet the dietary needs of swordtails. Include high-quality flake or pellet food as a staple diet and supplement with live or frozen foods such as bloodworms, brine shrimp, and daphnia. Additionally, offer blanched vegetables like spinach or zucchini for added variety and nutrition.

Maintain Proper Lighting: Ensure your aquarium has appropriate lighting to mimic natural day and night cycles. Provide a photoperiod of around 8 to 10 hours of light per day, followed by a period of darkness. Avoid excessive or inadequate lighting, as it can stress swordtails and affect their behavior.

Monitor Tank Mates: Keep an eye on the interactions between swordtails and other tank mates. Remove any aggressive or incompatible tank mates that may harass or stress your swordtails. Choose peaceful tank mates that won't compete for resources or pose a threat to your swordtails.

Offer Enrichment: Stimulate natural behaviors and provide mental stimulation for your swordtails by adding enrichment to the aquarium. This can include adding floating plants, providing toys or aquarium-safe decorations for exploration, and arranging the tank layout to create interesting hiding spots and territories.

Observe and Interact: Spend time observing your swordtails' behavior and interactions daily. Get to know their individual personalities and habits. Interact with your swordtails by offering them food or playing with them using safe aquarium-safe toys or objects.

Maintain Ideal Temperature: Keep the water temperature within the recommended range of 72°F to 82°F (22°C to 28°C) for swordtails. Avoid drastic temperature fluctuations, as they can stress swordtails and make them more susceptible to illness.

Minimize Stress: Minimize stressors in the aquarium by avoiding sudden changes in water parameters, handling fish gently during maintenance, and providing a stable and secure environment. Stress weakens the immune system and makes swordtails more vulnerable to diseases and health problems.

By following these tips and providing proper care and attention, you can enhance the well-being and quality of life for your swordtail fish, leading to happy, healthy, and vibrant fish in your aquarium.

CONCLUSION

In conclusion, swordtail fish (Xiphophorus hellerii) make fantastic additions to freshwater aquariums due to their vibrant colors, active behavior, and ease of care. Throughout this guide, we've covered various aspects of swordtail fish care, including their history, characteristics, tank setup, feeding, breeding, and more.

By providing a suitable environment with proper water parameters, adequate space, hiding spots, and a balanced diet, you can ensure the health and well-being of your swordtails. Additionally, selecting compatible tank mates and monitoring water quality are crucial for maintaining a harmonious community tank.

Remember to observe your swordtails regularly, interact with them, and address any signs of illness or stress promptly. With proper care and attention, swordtail fish can thrive and bring joy to any aquarium enthusiast.

We hope this guide has been informative and helpful in your journey to becoming a successful swordtail fish keeper. If you have any further questions or need assistance, don't hesitate to seek advice from reputable sources or experienced aquarists. Enjoy your journey into the fascinating world of swordtail fish keeping!

www.ingramcontent.com/pod-product-compliance
Lightning Source LLC
Chambersburg PA
CBHW050314230526
45471CB00005B/2184